UNDERWATER

DIVE

COLLINS

ISBN 0-00-196135-7
Copyright © 1989 Victoria
House Limited.
All rights reserved.
First published in the UK
by William Collins Sons and
Company Ltd, 8 Grafton
Street, London W1X 3LA.
Printed in the UK.

UNDERWATER

DIVE

Written by Dr. Antony Jensen and Dr. Stephen Bolt
Illustrated by Paul Johnson
Edited by Moira Butterfield

Contents

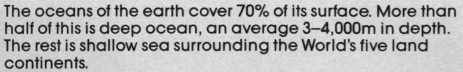

The oceans of the earth cover 70% of its surface. More than half of this is deep ocean, an average 3–4,000m in depth. The rest is shallow sea surrounding the World's five land continents.

Diving expeditions are the best way to study this enormous area of the planet, much of which is still unexplored. Wildlife can be found in nearly every area of ocean, so there is lots to see and learn.

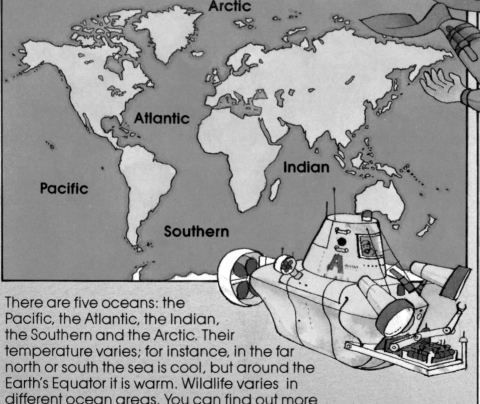

There are five oceans: the Pacific, the Atlantic, the Indian, the Southern and the Arctic. Their temperature varies; for instance, in the far north or south the sea is cool, but around the Earth's Equator it is warm. Wildlife varies in different ocean areas. You can find out more about this on page 12.

The depth and size of the oceans make them difficult and expensive to study, so only small sections have so far been explored. Underwater vehicles called submersibles are needed to reach the deepest parts (see page 8).

Since the invention of the aqualung in the 1940s divers have been able to explore the World's shallow seas.

An aqualung is a cylinder of compressed air carried on the back (see page 6). It gives divers the freedom to move about wherever they want to in shallow water.

Shallow seas are often the most interesting diving sites because they contain a rich variety of animals and plants, from warm water coral reefs to cool water seaweed forests. You can find out more about different diving sites on page 12.

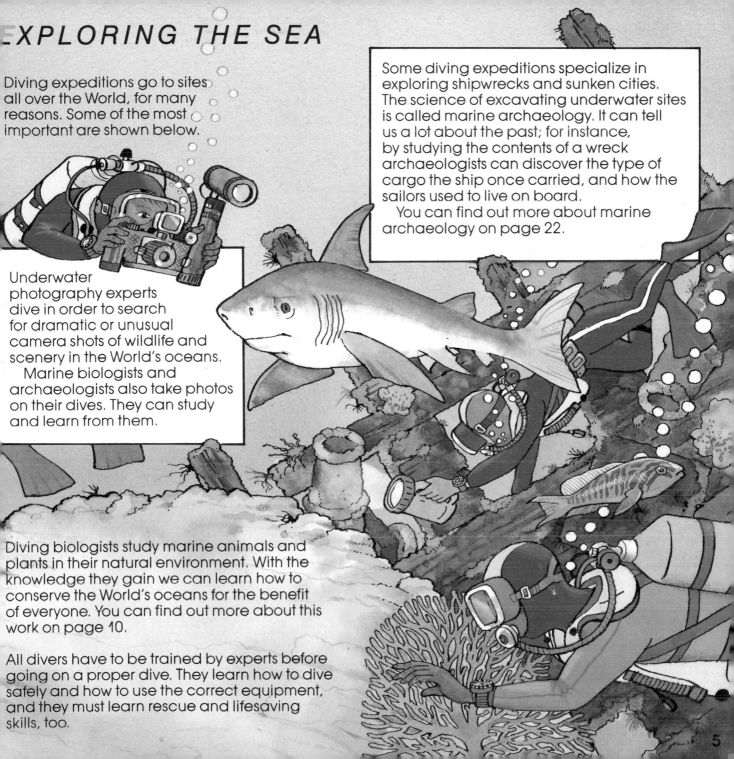

EXPLORING THE SEA

Diving expeditions go to sites all over the World, for many reasons. Some of the most important are shown below.

Some diving expeditions specialize in exploring shipwrecks and sunken cities. The science of excavating underwater sites is called marine archaeology. It can tell us a lot about the past; for instance, by studying the contents of a wreck archaeologists can discover the type of cargo the ship once carried, and how the sailors used to live on board.

You can find out more about marine archaeology on page 22.

Underwater photography experts dive in order to search for dramatic or unusual camera shots of wildlife and scenery in the World's oceans.

Marine biologists and archaeologists also take photos on their dives. They can study and learn from them.

Diving biologists study marine animals and plants in their natural environment. With the knowledge they gain we can learn how to conserve the World's oceans for the benefit of everyone. You can find out more about this work on page 10.

All divers have to be trained by experts before going on a proper dive. They learn how to dive safely and how to use the correct equipment, and they must learn rescue and lifesaving skills, too.

Diving expeditions have to be planned very carefully in advance, and all the underwater equipment has to be checked thoroughly before use, to make sure it works properly. Some of the vital equipment you need for a dive is shown below.

A diving suit helps you to keep warm and will protect you from sharp rocks and stinging creatures.

In very cold water you need a waterproof 'dry suit', so that you can put warm woollen underwear on beneath it.

In warmer water you can wear a 'wet suit', which lets some water through to the skin. This moisture is heated by the body and helps to keep you warm.

Around your waist you need a belt with heavy lead weights attached, to help you swim downwards. A pair of fins will help you to swim more strongly.

A face mask allows you to see clearly underwater. Everything you see through the mask will be magnified by a third, so animals and plants will look bigger than they really are.

When you dive, the water around you presses in on your body. The deeper you go, the higher the water pressure.

In order for your lungs to expand properly, the air you breathe must push outwards with the same pressure as the water pushes inwards. For this reason you need an aqualung.

Part of the aqualung is a cylinder filled with compressed air, which means that the air molecules are squeezed together so that the air exerts a high pressure outwards.

The 'demand valve' or 'regulator' part of the aqualung supplies the air at the same pressure as the water around you, however deep.

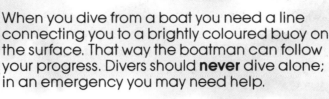

When you dive from a boat you need a line connecting you to a brightly coloured buoy on the surface. That way the boatman can follow your progress. Divers should **never** dive alone; in an emergency you may need help.

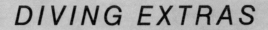

A knife is essential for cutting yourself free from any weeds or fishing nets.

A snorkel tube allows you to breathe on the surface.

You need an underwater torch at night or in murky water.

A divers' lifejacket keeps you buoyant on the surface, and helps you float up in an emergency.

A depth gauge is vital to check how deep you are.

Underwater cameras need a waterproof case and a flash gun.

When you breathe in air, the gases oxygen and nitrogen dissolve into your blood.

As you go below the surface, the pressure of the water makes more nitrogen dissolve than normal, and the deeper you go and the longer you stay there, the more nitrogen you absorb.

If you then surface too quickly, the nitrogen will form bubbles in your bloodstream, a condition called 'the bends'.

To avoid this you must look at 'diving tables', which help you calculate how deep and how long your dive should be. You must wear an underwater watch to time yourself and a depth gauge to measure how deep you are.

To take rock samples you need a hammer and chisel. A water-proof board and pencil can be used for notes.

After a dive you will need to shower in fresh water. Otherwise you could get salt-water skin sores. Antiseptic creams and powders should be added to the equipment list to treat these.

DEEP DIVING EQUIPMENT

Humans cannot dive in deep ocean because the very high water pressure soon disrupts the body's nervous system. Underwater vehicles called submersibles are used instead, for jobs such as repairing pipelines and salvaging wrecks. A submersible must be launched from a mother ship, with expert back-up crew and equipment on board to track the sub's movements.

Manned submersibles take between one and three people. They run on batteries and can stay underwater for a few days at a time.

Inside the sub there must be a life-support system for the crew, so that they can breathe and work comfortably.

Outside a sub there are usually observation windows, video cameras, spotlights and remote-controlled grabbing arms for picking up samples or manipulating tools.

The suit shown above is a one-man submersible called JIM. It can take someone down 300m.

Inside the suit there is a life-support system and outside there are mechanical hand-grabs and joints to allow for movement.

The JIM suit is useful for getting into confined spaces where bigger submersibles cannot reach.

Scientists use ROVs and submersibles to map the sea bed and to study deep ocean wildlife, sometimes putting out bait to attract fish near to the sub so that they can be filmed.

You can find out more about some of the deep-sea creatures seen from subs on page 18.

Unmanned submersibles are called ROVs (remotely operated vehicles). They usually have a cable linking them to the mother ship, so that someone on board can steer and remotely-control any robot arms.

Video cameras transmit the underwater pictures up to the surface. In deep, murky waters ordinary cameras are useless, so ROVs are fitted with 'low light intensity' cameras, which are sensitive to small amounts of light.

If a diver gets ' the bends ' by going too deep he must be rushed into a decompression chamber, where the air pressure is increased to dissolve nitrogren bubbles back into the blood. Then the pressure is brought slowly back to normal.

When divers are working constantly at great depths their bodies become 'saturated', absorbing all the nitrogen they can. It is most convenient for them to go up and down each day in a pressurized diving bell, spending their off-duty time in a pressurized chamber. When their job is finished they go into a decompression chamber to correct their nitrogen levels.

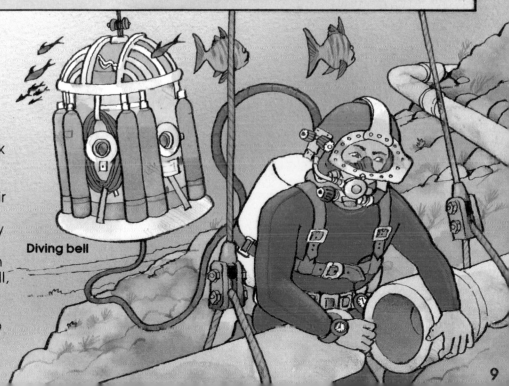

Diving bell

STUDYING THE SEA

Divers play a valuable part in increasing our knowledge of the oceans because they can study marine life at first-hand, watching creatures in their natural habitat.

It can take years to plan an expedition to study wildlife. First, diving scientists must decide exactly what they want to study — for instance, why a coral reef is being damaged, or how pollution is affecting an area of sea-bed.

You don't need to be a trained scientist to help with this work. For example, if biologists need to know where animals and plants live around a coastline, amateur divers can help by collecting samples and observing wildlife.

Waterproof boards and pencils are useful for recording underwater sights, and plastic bags are used for taking samples to the surface for further study.

Photographs of underwater life are very valuable to marine scientists, who can study them in detail later.

Because visibility may be poor you usually have to get as close as possible to a subject and use a wide angle lens to get as much of it in the picture as you can.

Once collected, wildlife samples must be preserved and labelled with their name and where they were found. Months of laboratory research often follow.

One example of underwater biology work is the study of coral reefs. Some fascinating scientific experiments have been done on them. For instance, by placing different sorts of coral together divers have discovered that some types are very aggressive, killing their neighbours by giving off destructive chemicals.

This type of experiment takes a long time, and can mean dives to the same site for months or even years.

The best way to study fish is in their natural environment, the ocean. But there are problems in trying to watch them underwater because there is a chance you might disturb them and frighten them away.

Fish can sense water pressure and flow changes caused by other creatures some distance away. They do this through a system of sensitive canals and openings along their body, called the 'lateral line'.

Lateral line

Diving geologists do important scientific work, too. They explore areas marked out for tunnelling or oil exploration, taking samples of rocks or sea-bed and bringing them to the surface for study.

DIVERS' SEALIFE GUIDE

A wildlife community living in one particular area is called an ecosystem.

Ocean ecosystems vary according to sea conditions such as depth and temperature.

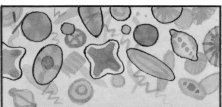

In all ecosystems there must be a food chain. The basic ocean food source is algae, a tiny floating plant which gets its nutrients (nourishment) from the seawater around it.

Algae are eaten by fish, whales and tiny animals called plankton. But many sea creatures depend on each other for food, hunting and attacking other ocean animals.

At the end of the food chain there are sea-bed creatures which eat dead material. When they digest this food they break it down into nutrients which are recycled into the sea.

ROCKY SEA-BED

If you dive in a rocky sea-bed area you are likely to see lots of varied wildlife. A rock is a stable, solid surface for plants and animals to live on, whilst cracks and caves provide plenty of nooks and crannies for animals to set up home and hide from their enemies.

FLAT SEA-BED

Much of the sea-bed is flat and featureless. However, that does not stop animals living there. For example, many creatures live in mud, burrowing down to hide from their enemies. Sea cucumbers, worms and crabs are common bottom-dwellers.

MID-OCEAN

Most ocean wildlife floats or swims near the sea surface in mid-ocean, far from the shore or sea-bed. Many fish and jellyfish never venture near the shore or the sea-bottom.

CORAL REEFS

Coral reefs are ideal for divers interested in wildlife. They provide a home for many creatures.

The water around reefs is usually clear and good for photography. Even the sounds made by underwater creatures can sometimes be heard.

POLAR REGIONS

If you went diving under polar ice you would see lots of creatures living in the freezing water. Millions of tiny shrimp-like creatures called amphipods live upside-down on the ice undersurface. Fish eat the amphipods and creatures such as seals eat the fish as part of the polar food chain.

DEEP SEA

It is difficult to explore the deepest parts of the sea. Many areas are too deep for sunlight to reach, so the creatures who live there have developed features such as luminous spots to help them feed. You can find out more about this strange ecosystem on page 18.

THE SEA GARDEN

In shallow seas the underwater landscape can look like a spectacular garden, made up of seaweeds and plant-like animals such as corals and anemones.

1/ Corals are colonies of animals which live in warm, clear water. They are either hard or soft, and they grow in many shapes; for instance, some look like fingers, others like cabbage leaves or delicate fan-shaped skeletons.

2/ Soft corals have an outside layer of living tissue surrounding a central rod which can bend in currents.

The tissue is home to many tiny anemones called polyps, connected to each other through their stomachs. They use tiny tentacles to catch food from the water, which they share between them.

3/ Large corals are usually hard. They have a 'wall' around them which they build by taking calcium from the water and using it to surround each of the many little polyps that live in the coral tissue.

As the polyps grow and increase in number the size of the coral increases, and eventually reefs are formed.

4/ Brightly-coloured sponges and anemones are common coral reef animals, and many vividly-marked fish are attracted to corals, often forming spectacular shoals around them.

Some animals damage reefs badly. For instance, the crown-of-thorns starfish feeds on coral polyps in warm waters.

Crown-of-thorns

Behind a reef there is often a shallow lagoon carpeted with sea grasses, the only true flowering plants in the ocean.

Sea grass meadows are nursery grounds for fish and feeding grounds for animals such as turtles. In some meadows you might see the dugong ('sea cow'), a strange-looking and gentle-natured marine mammal.

Dugong

The kelp seaweed forests of the eastern Pacific are a spectacular diving sight, with seaweed towering up to 50m high. The forests are famous for sea otters, which help the seaweed to survive by eating the urchins that graze on young kelp plants.

Sea otters

The otters dive to the sea-bed, pick up an urchin and a stone and take them both up to the surface. They float on their backs and, using their chests as tables, they break open the urchin with the stone and eat the contents.

More beauty lies beneath the kelp blades, where sponges, anemones and starfish provide brilliant colours for divers to see.

DANGEROUS ANIMALS

1/ Amongst the many thousands of creatures in the sea there are a few which are harmful, sometimes even fatal, to humans.

There are two sorts of dangerous creatures divers need to look out for — the kind that bite and the kind that sting.

4/ Sharks are the largest sea fish. Some species feed peacefully on plankton, but most sharks feed on other creatures and have powerful jaws filled with teeth.

Not all sharks are dangerous, but a few species will attack humans. Divers need t look out particularly for the hammerhead shark, shown below, and the great white shark, which can grow up to 4.6m long.

Shark attacks on divers are quite rare, but you would increase the chances by spear fishing, as blood from a harpooned fish would attract hungry sharks. An attack is also more likely if you splash around on the water surface. The sharks could mistake you for wounded prey.

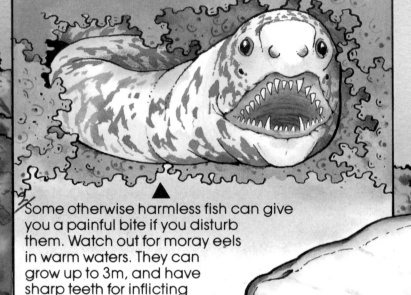

2/ Some otherwise harmless fish can give you a painful bite if you disturb them. Watch out for moray eels in warm waters. They can grow up to 3m, and have sharp teeth for inflicting nasty wounds.

3/ In many areas sea snakes are common; in some cases their venom is more poisonous than any land snake. Luckily, they are not usually aggressive and will leave you alone unless they are disturbed or frightened by sudden movements.

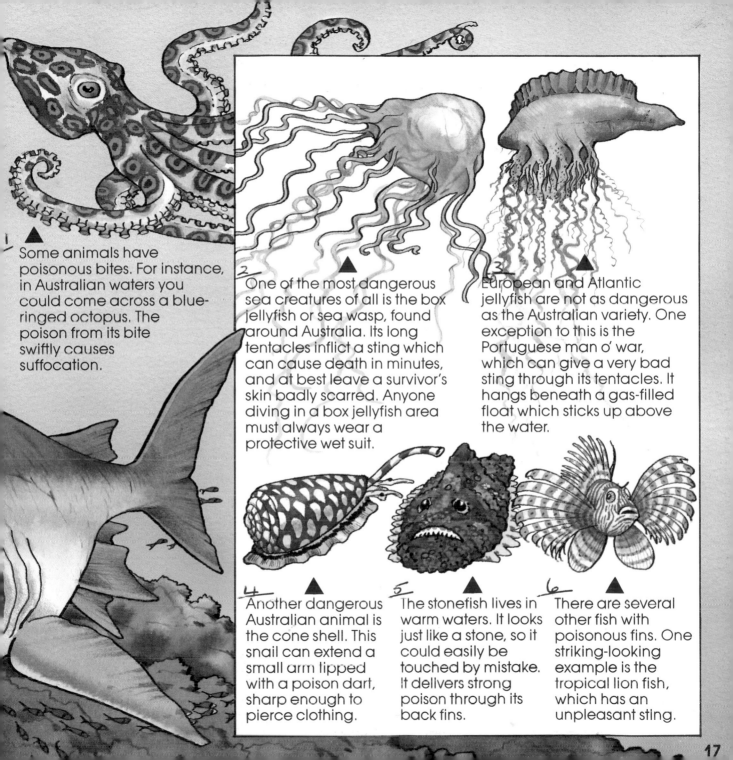

1 ▲
Some animals have poisonous bites. For instance, in Australian waters you could come across a blue-ringed octopus. The poison from its bite swiftly causes suffocation.

2 ▲
One of the most dangerous sea creatures of all is the box jellyfish or sea wasp, found around Australia. Its long tentacles inflict a sting which can cause death in minutes, and at best leave a survivor's skin badly scarred. Anyone diving in a box jellyfish area must always wear a protective wet suit.

3 ▲
European and Atlantic jellyfish are not as dangerous as the Australian variety. One exception to this is the Portuguese man o' war, which can give a very bad sting through its tentacles. It hangs beneath a gas-filled float which sticks up above the water.

4 ▲
Another dangerous Australian animal is the cone shell. This snail can extend a small arm lipped with a poison dart, sharp enough to pierce clothing.

5 ▲
The stonefish lives in warm waters. It looks just like a stone, so it could easily be touched by mistake. It delivers strong poison through its back fins.

6 ▲
There are several other fish with poisonous fins. One striking-looking example is the tropical lion fish, which has an unpleasant sting.

DEEP-SEA LIFE

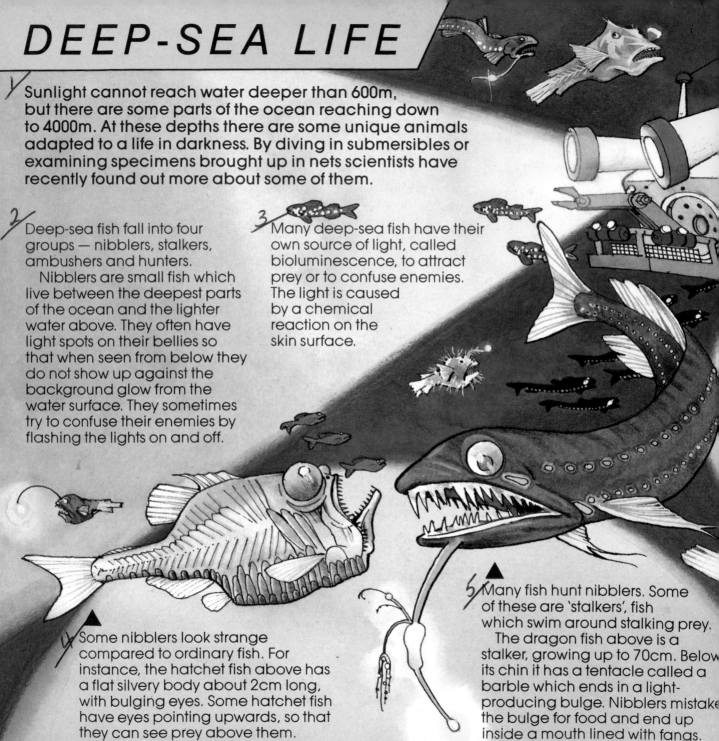

1 Sunlight cannot reach water deeper than 600m, but there are some parts of the ocean reaching down to 4000m. At these depths there are some unique animals adapted to a life in darkness. By diving in submersibles or examining specimens brought up in nets scientists have recently found out more about some of them.

2 Deep-sea fish fall into four groups — nibblers, stalkers, ambushers and hunters.

Nibblers are small fish which live between the deepest parts of the ocean and the lighter water above. They often have light spots on their bellies so that when seen from below they do not show up against the background glow from the water surface. They sometimes try to confuse their enemies by flashing the lights on and off.

3 Many deep-sea fish have their own source of light, called bioluminescence, to attract prey or to confuse enemies. The light is caused by a chemical reaction on the skin surface.

4 Some nibblers look strange compared to ordinary fish. For instance, the hatchet fish above has a flat silvery body about 2cm long, with bulging eyes. Some hatchet fish have eyes pointing upwards, so that they can see prey above them.

5 Many fish hunt nibblers. Some of these are 'stalkers', fish which swim around stalking prey. The dragon fish above is a stalker, growing up to 70cm. Below its chin it has a tentacle called a barble which ends in a light-producing bulge. Nibblers mistake the bulge for food and end up inside a mouth lined with fangs.

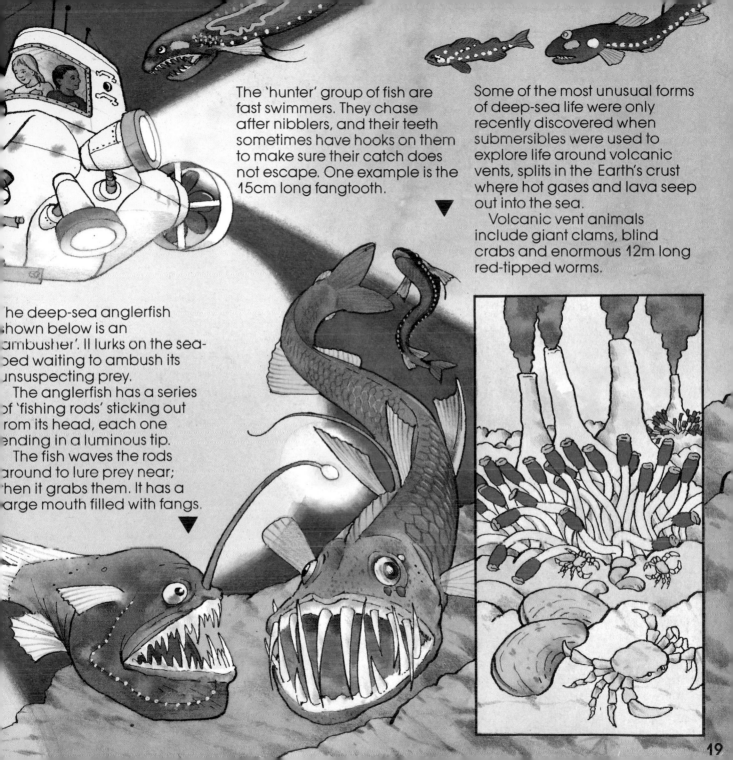

The 'hunter' group of fish are fast swimmers. They chase after nibblers, and their teeth sometimes have hooks on them to make sure their catch does not escape. One example is the 15cm long fangtooth.

Some of the most unusual forms of deep-sea life were only recently discovered when submersibles were used to explore life around volcanic vents, splits in the Earth's crust where hot gases and lava seep out into the sea.

Volcanic vent animals include giant clams, blind crabs and enormous 12m long red-tipped worms.

The deep-sea anglerfish shown below is an 'ambusher'. It lurks on the sea-bed waiting to ambush its unsuspecting prey.

The anglerfish has a series of 'fishing rods' sticking out from its head, each one ending in a luminous tip.

The fish waves the rods around to lure prey near; then it grabs them. It has a large mouth filled with fangs.

19

Of all the animals in the oceans, whales, dolphins and porpoises are most closely related to humans. Like us, they are mammals with warm blood. Most other fish have cold blood. These sea mammals are the most intelligent creatures in the sea, and among the most interesting for divers to study.

There are two types of whale — 'toothed' and 'baleen'. Instead of teeth baleen whales have horny plates which filter tiny floating animals called krill from the seawater. The biggest animal in the world, the blue whale, is a baleen whale, growing up to 33m long. Toothed whales, such as sperm whales, eat larger fish and squid.

You can tell dolphins and porpoises apart by their shape. Porpoises have blunt snouts and fat bodies; dolphins are bigger, with longer snouts.

Dolphin

Blue whale

Dolphins and porpoises are very popular with divers because of their playful nature. Diving alongside whales is much more difficult. Observation has been limited to brief encounters near the water surface, when whales will sometimes tolerate humans coming near.

Despite this, diving is the best way to study whales at close hand and there have been many useful expeditions. One example was a diving expedition organised to observe sperm whales in the Indian Ocean. The divers discovered a lot about the behaviour of the whales and were the first to see a sperm whale born.

There are still some sea mammal mysteries which divers could help to solve in the future. For example, whales can dive to great depths for long periods of time but we do not yet know how their bodies can stand up to the high water pressure at deep levels.

Whales, dolphins and porpoises make noises such as clicks, barks and whistles. When they are in groups they use the noises to 'talk' to each other, but no-one has yet been able to understand their language.

Sea mammals also use noise to detect objects in darkness. To do this they make a sound and then detect the returning vibrations bouncing off objects in their path.

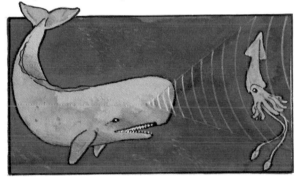

Whales are hunted for use in many commercial products. This has led to the near extinction of several species, and now conservationists are trying to limit the killing before it is too late.

Divers have helped in this battle by studying and filming whales. In this way more people have become aware of their beauty and intelligence, and the need to save these giants of the ocean.

Porpoise

HISTORY SEARCH

Diving archaeologists explore the remains of ships or sunken towns. As in land archaeology, an excavation underwater is called a 'dig'.

It can take years to excavate a dig, with a lot of painstaking planning and hard work in the water and on the surface, looking after equipment and writing reports on the finds of the day.

A survey is made of the dig site, and sometimes a large metal grid is placed over the area. This can be used like a map for reference and divers can swing from it as they work. They use metal detectors and probes to explore the sea-bed.

It is not easy to find a wrecked ship. If it sank on rocks the hull might have broken up and spread over a wide area. If it landed on a sandy or muddy sea-bed it could be buried in sludge.

Archaeologists searching for a particular ship will look through many library archives, reading different stories about where and how a ship sank and gathering clues on its location.

To pinpoint wrecks from the surface archaeologists use magnetometers, which detect the magnetism of metal objects. They also use sonars, bouncing sound waves off the sea-bed and measuring the time they take to return. This data is used to build up a sea-bed picture.

To clear away mud and sand divers use an air lift, a long pipe that has air pumped up into the end nearest the sea-bed. As the air rises up the pipe it sucks up mud and sand as well. The flow of air can be altered to excavate small objects or large areas.

Every object found on an archaeological dig is valuable as a possible historical clue. Each piece must be carefully cleaned, drawn and photographed underwater, and it is very important to record exactly *where* each object is found on a wreck site. This helps archaeologists decide how things were used and who might have owned them; for example, if an object comes from a ship's galley it could well have been used in cooking.

Finds are very gently lifted to the surface for careful restoration. This can take some time; for instance, wood has to be washed in fresh water for many days to remove the sea salt. Then it is soaked in a chemical to stop it splitting and rotting. Conserving a very large wooden object, such as a ship's hull, can take up to 20 years.

Some objects last very well underwater. One example is this barber-surgeon's chest from the *Mary Rose*, which sank off Britain in 1545. Inside were ointment jars, black pepper and medical tools.

DIVING HISTORY

Although modern diving equipment was not developed until the 1930s, the history of diving goes back a long way.

The 'turtle'

The first known divers were people who gathered shellfish to eat and items they could sell, such as sponges and pearls. They could not stay under water for long, so to speed up their descent they would tie themselves to lumps of stone.

From ancient Greek times divers have also been used to salvage treasure from sunken ships and to sabotage the enemy in sea battles.

We know that diving equipment was used from Roman times, when divers used bamboo snorkels to swim below the surface.

From the 1300s onwards inventors started designing more complicated equipment. Even the great artist Leonardo Da Vinci did some diving sketches in the 1500s. Most of these early ideas could never have worked in practice.

The wooden 'turtle' was the first submarine design to work. It was built by an American called David Bushnell, and was used to attack the British Fleet off the American coast in 1776.

Another early submarine was the American Nautilus of 1801. It carried three people, one as Captain and two to turn the propeller by hand. It had a sail which folded up like an umbrella when underwater.

One of Da Vinci's ideas

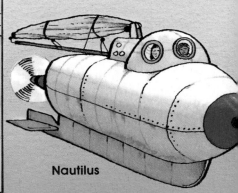

Nautilus

Once efficient air pumps were invented diving suits became possible. They usually had heavy glass-fronted helmets and were sometimes called 'hard hat' suits.

In the 1800s there were lots of new designs culminating in the Siebe suit, shown below.

For safety the whole of the Siebe suit was inflated with air. This made the diver very light, so he had to wear heavy boots and chest weights.

'Hard hat' divers needed an air-line going up to the surface. The first person to think of designing a self-contained suit was the American W.H. James. His idea was to carry an air supply inside a metal belt, as shown on the left. A similar suit was used by Charles Condert in New York's East River. He died whilst diving in 1832, when an air tube broke.

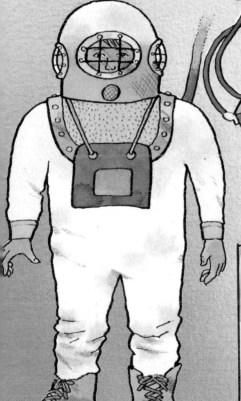

This 1808 suit design would never have worked. The diver wears a crown attached to some bellows, which he works by nodding.

This strange and impossible set-up was designed in 1551. The diver is in a weighted wooden frame, with his head inside a glass ball.

This is an early design for a pressurized air cylinder, pioneered by the Frenchmen Rouquayrol and Denayrouze.

MYTHS AND LEGENDS

The sea has for many centuries been the subject of incredible tales and mysterious happenings. The black depths would make perfect homes for monsters and magical underwater kingdoms.
   Modern divers have been able to explain many of the old stories, but there are still a few which remain mysteries.

Half-fish, half-human mermaids are a common sea legend. In many stories they come ashore as women, marry humans and later disappear back to the sea. In Greek legend beautiful maidens called sirens lured ships onto rocks.
   These legends may have come originally from sightings of seals and dugongs (see page 15). In most mermaid stories the moral is that strange sea creatures should be left well alone!

The sailors of the past feared the attack of giant sea monsters. Pictures of them can be seen on many ancient sea-maps. To sailors in a small boat a spouting whale could have been mistaken for a monster.

One 'sea monster' that does exist is the Atlantic giant squid, and there are true reports of these squids attacking small boats. The longest squid ever measured was over 17m long, with tentacles that measured another 15m.

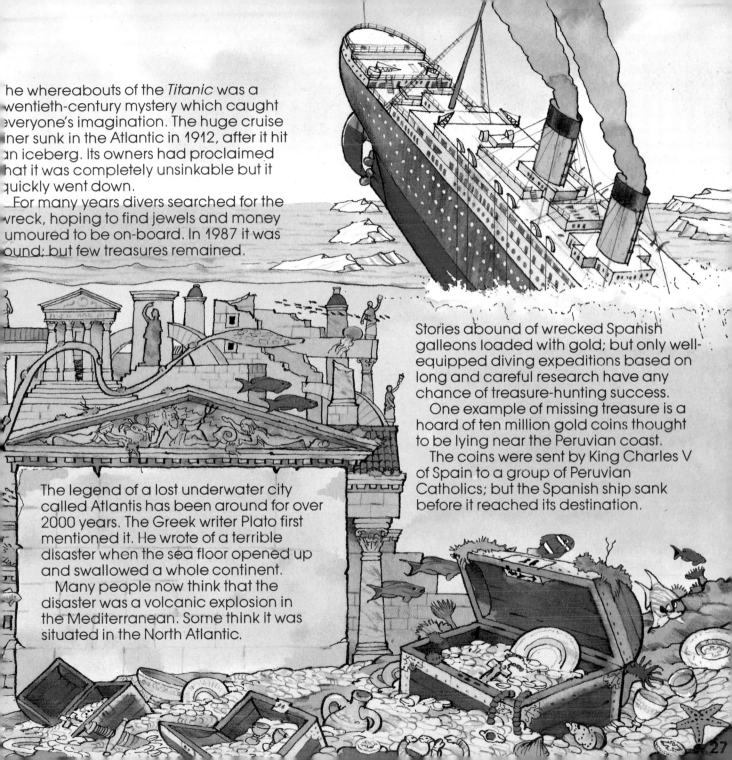

he whereabouts of the *Titanic* was a wentieth-century mystery which caught everyone's imagination. The huge cruise iner sunk in the Atlantic in 1912, after it hit an iceberg. Its owners had proclaimed hat it was completely unsinkable but it quickly went down.

For many years divers searched for the wreck, hoping to find jewels and money umoured to be on-board. In 1987 it was ound; but few treasures remained.

The legend of a lost underwater city called Atlantis has been around for over 2000 years. The Greek writer Plato first mentioned it. He wrote of a terrible disaster when the sea floor opened up and swallowed a whole continent.

Many people now think that the disaster was a volcanic explosion in the Mediterranean. Some think it was situated in the North Atlantic.

Stories abound of wrecked Spanish galleons loaded with gold; but only well-equipped diving expeditions based on long and careful research have any chance of treasure-hunting success.

One example of missing treasure is a hoard of ten million gold coins thought to be lying near the Peruvian coast.

The coins were sent by King Charles V of Spain to a group of Peruvian Catholics; but the Spanish ship sank before it reached its destination.

UNDERWATER FUTURE

Future scientific developments will probably lead to safer and easier diving in the World's oceans.

Work is now being done on lighter, more comfortable diving suits and aqualungs capable of carrying more breathing air.

New materials could be developed for dry suits (see page 6). These could insulate against the cold more efficiently and make the suit less buoyant so divers will need fewer weights on their belts.

These suits of the future will probably have heating pads to keep the diver warm, and better communications equipment.

Small waterproof decompression computers will probably become essential equipment, making it easier for divers to work out how long they should stay underwater and how slowly they should come up to the surface to avoid the bends (see page 7).

In the 1960s the French diving pioneer Jacques Cousteau built an underwater house called 'Conshelf', to find out whether humans could live underwater.

Inside Conshelf the air was kept at the same pressure as the water outside. This caused problems for the divers who stayed there, and Conshelf was finally closed.

To avoid these difficulties underwater houses could possibly be built in shallow water, where inside pressure could be kept at a level similar to the air pressure above the surface. The inhabitants could stay under for as long as they wanted. For deeper underwater work they would need armoured diving suits or submersibles pressurized at the same level as their houses.

arming of the sea is called mariculture. Several different types of animal and plant are farmed, including oysters, salmon and seaweed.

In the future, mariculture is likely to become even more important. Already scientists have managed to build artificial reefs and 'seed' them with lobsters and other shellfish. Special breeds of fish may be developed that are easier to keep in captivity and are resistant to diseases and parasites. Small underwater communities could be built for the marine farm workers.

FUTURE PARKS

As sports diving has become more popular, wildlife has sometimes been threatened. A few years ago, divers would commonly take lobsters and crabs and use spear guns to kill fish.

More recently underwater areas have been made into marine reserves, where divers take nothing except photographs, just like the massive game parks of Africa where big game hunting once threatened to ruin wildlife. One of those reserves, the Coral Reef State Park in Florida, is dotted with beautiful underwater statues, such as the one shown on the left.

As well as making sure that sea life is preserved, these areas give scientific divers a great opportunity to study underwater creatures in their natural habitat. The reserves will eventually provide a source of young animals to restock areas of the sea.

INDEX

History of diving picture reference –
Sir Robert Davis, Deep diving and
submarine operations: a manual for
deep-sea divers and compressed air
workers. 1962.

If you want to know more about diving
and ocean life you can contact the
addresses shown below:

Marine Conservation Society,
9, Gloucester Place,
Ross-on-Wye HR9 5BU
Tel: 0989 66017

Greenpeace,
30–31, Islington Green,
London NW 8XE
Tel: 01-354 5100